AF791101

SPECTACULAR
AFRICAN WILDLIFE

Struik Publishers (Pty) Ltd
(a member of the Struik New Holland Publishing Group (Pty) Ltd)
80 McKenzie Street
Cape Town 8001

Originally published in 1998 as *Magic of African Wildlife*

Reg. No.: 1954/000965/07

First edition 1999
3 5 7 9 10 8 6 4 2

Project manager: Pippa Parker
Editor: Peter Joyce
Designer: Dominic Robson

Reproduction by Hirt and Carter (Pty) Ltd, Cape Town
Printed and bound by Kyodo Printing Co (S'Pore) Pte Ltd

ISBN 1 86872 483 2

Front Cover (top) : Leopard *(Lex Hes)*
(bottom, left): Chimpanzee *(Martin Harvey)*
(bottom, right): Spotted hyaena *(Nigel Dennis/ABPL)*
Back cover: Lions at watering hole *(Nigel Dennis)*
Title page: Springbok *(Hein von Hörsten/SIL)*
Below: Black-backed jackal *(Daryl Balfour)*
Opposite page: Burchell's Zebra *(Peter Pickford/SIL)*

Contributing photographers include Shaen Adey, Daryl Balfour, Adrian Bailey, Roger de la Harpe, Nigel Dennis, Richard du Toit, Albert Froneman, Lex Hes, Hein von Hörsten, Walter Knirr, Colin Paterson-Jones, Peter Pickford, Warwick Tarboton, Pieter Viljoen and Keith Young.

MAJOR WILDLIFE RESERVES OF AFRICA

1. Banc D'Arguin National Park
2. Delta du Saloum National Park
3. Boucle du Baoule National Park
4. Niokolo-Koba National Park
5. Badiar National Park
6. Outamba-Kilimi National Park
7. Sapo National Park
8. Mont Peko National Park
9. Comoe National Park
10. Mole National Park
11. Digya National Park
12. Old Oyo National Park
13. Zakouma National Park
14. Dinder National Park
15. Simien Mountains National Park
16. Bamingui-Bangoran National Park
17. Korup National Park
18. Odzala National Park
19. Garamba National Park
20. Maiko National Park
21. Salonga National Park
22. Virunga National Park
23. Kundelungu National Park
24. Lusenga Plain National Park
25. Kidepo National Park
26. Kabalega Falls National Park
27. Sibiloi National Park
28. Marsabit National Reserve
29. Samburu National Reserve
30. Mount Elgon National Park
31. Tsavo National Park
32. Serengeti National Park
33. Gombe Stream National Park
34. Ruaha National Park
35. Selous Game Reserve
36. Nyika National Park
37. North and South Luangwa National Park
38. Kafue National Park
39. Kangandala National Park
40. Kisama National Park
41. Bikuar National Park
42. Iona National Park
43. Mupa National Park
44. Etosha National Park
45. Cape Cross Seal Reserve
46. Namib-Naukluft National Park
47. Moremi Wildlife Reserve
48. Chobe National Park
49. Gemsbok National Park
50. Mana Pools National Park
51. Hwange National Park
52. Pilanesberg National Park
53. Kruger National Park
54. Mkuzi Game Reserve
55. Greater St Lucia Wetland Park
56. Hluhluwe-Umfolozi Park
57. Itala Game Reserve
58. Golden Gate Highlands National Park
59. Royal Natal National Park
60. Addo Elephant National Park
61. Mountain Zebra National Park
62. Zuurberg National Park
63. Wilderness National Park
64. Karoo National Park
65. Augrabies Falls National Park
66. Kalahari Gemsbok National Park
67. Richtersveld National Park
68. Goegap Nature Reserve
69. West Coast National Park
70. Cape of Good Hope Nature Reserve
71. De Hoop Nature Reserve

Introduction

Africa is the second largest (after Eurasia) of the world's continents, a gigantic landmass that embraces many different climates, a myriad landscapes and a wide range of natural habitats that, between them, encompass a breathtaking diversity of life forms. No place on earth bears such clear and continuing witness to the variety, the beauty and the drama of the wild kingdom.

Much of the north is desert, though it has not always been so. Just a few millennia ago the retreating chill of the last ice age left behind a land that was fertile, supporting wooded savannas, grasslands, great numbers of wild animals – and rich pastures. Indeed, the people who came to dominate large parts of the region, the Carthaginians, provided the Roman Empire with much of its food requirements. But the scene has changed over the intervening centuries: the air has become drier, rain rarely falls, and the deep soils have given way to stoney plains and high white dunes rolling away to horizons that shimmer in the searing heat. And the Sahara is advancing ever farther southwards: each year its southern fringes, known as the Sahel, become a little more arid, a little less able to sustain the nomadic pastoralists that call it home.

Beyond the Sahel are the rainforests of tropical Africa, a vast green mantle stretching from the Atlantic coast deep into the continent's heart. The forests here are under threat, gradually disappearing before the twin onslaught of natural pressures and human encroachment; many of the West African ones have been reduced to mere patches, and even these are shrinking as the land is cleared for settlement and cultivation, and as people move south from a Sahel that is becoming increasingly inhospitable. Nevertheless, much remains of the original, dense equatorial cover and its living organisms, especially in the central regions. The moist green depths of the Congo (formerly Zaire) forests, for instance, host more than 400 mammal species, nearly 1 100 different kinds of bird, some of them lovely to behold in their many-hued finery, and countless reptiles, insects and plant forms. Most of the animals are small and secretive, but not all: here you will find the duiker and the strange okapi (discovered relatively recently; it looks something like a cross between a giraffe and zebra), snuffling bushpigs, dwarf antelope, shy leopard and, largest of them all, the forest elephant. However, the most notable, certainly the best publicized, denizens are the primates: chimpanzees and their relatives the bonobos, various species of monkey and the three races of gorilla (eastern lowland, western lowland and mountain). All in all, the wider equatorial region is haven to some 50 different primates, the Congo itself to more than 30.

Entirely different in character and wildlife content is the more open countryside of East Africa, a land of lovely tropical coasts, broad grassland plains, rivers, lakes and highland landscapes sweeping up to heights that were born of stupendous volcanic eruptions and which stun the senses with their grandeur. Among the most prominent of these are Mount Kilimanjaro, at 5 896 metres above sea-level the highest of Africa's peaks; Mount Kenya, Mount Meru and, to the west, across the great divide that separates Lake

Top right: **The lion's massive paw – a potentially lethal weapon.** *(Dennis/SIL)*

Right: **Wild dogs** *Lycaon pictus* **rest up in the heat of an African afternoon.** *(Dennis/SIL)*

Victoria and the uplands of the east from the lush reaches of the Congo River, the legendary Mountains of the Moon. All these and much more are products of an ancient and continuing geophysical process that created the Great Rift Valley, a 6 400-kilometre crack in the continental plateau that stretches from Syria and the Dead Sea southwards through Ethiopia and East Africa to disappear just north of the Zambezi River.

The Rift is at its most visually dramatic in Kenya, where its walls rise almost sheer to form an escarpment of exceptional beauty. On the lower ground are some of Africa's finest and most famous game areas, among them the Masai Mara Reserve, the fabulous Ngorongoro Crater and the nearby plains of Serèngeti, across which two million and more herbivores – gazelle, zebra, topi, hartebeest and, most especially, wildebeest – make their seasonal migration to fresh pastures. This is perhaps the greatest wildlife show on earth, reaching its most dramatic moment when the close-packed, surging, jostling animals, advancing in columns that stretch back 40 kilometres and more, negotiate the fast-flowing Mara River. Many are drowned in the turbulent waters, providing a rare feast for the carnivores that follow and for the crocodiles that haunt the river banks.

As one travels south from the green luxuriance of the Congo, so the land becomes drier, the rains less predictable; forest gives way to well-wooded savanna and then to bush- and grasslands. Much of this is splendid safari country. The Luangwa River, one of the Zambezi's largest tributaries, rises in the Zambia-Malawi border area and, when the wet season is generous, will break its banks to inundate the surrounding countryside. Sometimes, after the floods, its course has changed to leave behind a scatter of ox-bow lakelets – a cycle that creates a marvellous diversity of landscape and habitat. It also provides a venue for one of the continent's largest and most impressive concentrations of game animals, among them elephants and buffalo herds.

Above: **The dainty Damara dik-dik** *Madoqua kirki,* **among the smallest of Africa's antelope, graces Namibia's dry woodlands.** *(Du Toit)*

Bottom: **The butterfly** *Aeropetes tulbaghia* **and its floral perch bring a bright splash of colour to the South African landscape.** *(Paterson-Jones)*

The Luangwa valley is one of several superlative southern African wildernesses, each with its distinctive geography, fauna and plant life, that beckon the traveller and nature lover. Not too far away is the Middle Zambezi valley, a rugged stretch of river, floodplain, alluvial terrace, woodland and high flanking escarpment that extends from the mighty Victoria Falls to the Mozambique border, and it also draws a huge concourse of wild animals. Upstream is Kariba, among the largest of man-made lakes, focus of a trio of national parks and a prime destination for the game-viewer.

Every one of the region's countries places conservation high on the list of priorities (for various reasons, not least for the part it plays in attracting free-spending tourists from abroad).

Zimbabwe's flagship is Hwange, once the royal hunting ground of Ndebele warrior-kings and now ranked among the aristocrats of

wildlife havens, its broad sandy acres hosting (in the dry season) more than 20 000 elephants. Botswana boasts the game-rich Chobe and Moremi reserves, the former renowned for its lions, the latter occupying a large segment of the Okavango swamps, a magical fantasia of floodplain and forested islands, streams, pools, lagoons, the occasional lake and a labyrinthine network of magical channels. This is one of the world's few inland deltas, and among the most splendid of its great wetland wildernesses.

Pride of the South African conservation effort is the famed Kruger National Park, a vast stretch of low-lying, heat-hazed bushveld countryside, bigger in extent than the state of Israel, that serves as home to more kinds of animal and bird life than any other sanctuary on the continent. The game areas of Zululand, on the country's eastern seaboard, are much less spacious but equally well endowed: the land here is well watered by rain and river, lush in its growth, and the warmth, the moisture and rich vegetation nurture a remarkable profusion of life forms (for instance one small portion of the region, the Hluhluwe section, is just one-twentieth the size of the Kruger yet contains 70 percent of the latter's species).

These are a few, a tiny cross-section, of the wonders of wild Africa. There is much more – an almost infinite variety of habitats and of the living organisms they support – to beguile the naturalist, the photographer and the game-spotter alike. Unhappily, though, there is nothing permanent about the wildernesses, and as each year goes by a little less remains unspoilt. Ecosystems are being degraded, species are disappearing at a tragic rate and many others are coming under threat as urban populations grow, as farm- and ranchlands encroach ever further over pristine terrain, and as poachers and 'collectors' go about their lucrative and highly destructive business.

Little enough, it seems, can be done to halt poaching and, in the short term at least, human populations will continue to expand at an alarming rate. The harsh reality, therefore, is that much of Africa's natural heritage is doomed.

But there is hope for the remainder. A new approach to conservation is gaining ground, one which makes provision for integrated reserves in which the people of the countryside, instead of being shut out, have access to the land, share in its resources, benefit from the funds generated by ecotourism and, in return, help preserve the environment. This is a situation in which everyone's interests (including those of the wildlife) are served, and it holds real promise for the future.

Top: A black eagle *Aquila verreauxii* **soars above the high cliffs of its mountain habitat.** *(Froneman)*

Below: The leopard *Panthera pardus* **is primarily a ground-living cat, but will take to the trees to rest, and to cache its kill.** *(Balfour)*

Provoked by human intrusion, an angry elephant *Loxodonta africana* **(opposite)** charges through the mud of a waterhole. Its ears serve not only to sharpen its hearing (and to display its irritation), but also to help cool its massive body in torrid weather. *(Du Toit)*

A redbilled oxpecker *Buphagus erythrorhynchus* **(below)** takes its ease on the muzzle of a buffalo, one of the 'big five' animals so sought after by the game-spotter. *(Du Toit)*

Animals two by two. A brace of suricates or meerkats *Suricata suricatta* **(above)** wrestle each other in mock battle. These endearing little creatures, which belong to the mongoose family, live in groups of up to 40 individuals. *(Dennis)*

Overleaf left: Burchell's zebra twosome. The bold markings of these animals might not seem to offer much in the way of camouflage, but the stripes do tend to distract predators: individuals are difficult to single out in the black-and-white confusion of the herd. *(Balfour)*

Overleaf right: The long neck of a giraffe *Giraffa camelopardalis*, tallest of all land mammals, enables the animal to reach leaves and shoots denied to other browsers. *(Balfour)*

Among the most resourceful and disciplined hunters of the African veld are wild dogs **(above)**. The rank structure within a pack is strictly defined, though it changes over the seasons. These two are involved in 'play' behaviour that helps establish the subtleties of seniority. *(Balfour)*

Male gemsbok *Oryx gazella* **(below)** fight each other for dominance over territory, and for the sole right to mate within the nursery herd. *(Du Toit)*

Previous pages: A white rhino *Ceratotherium simum* and her calf in retreat. These massive, primeval herbivores have been hunted to the edge of extinction for their horns, which are valued in the Far East for their reputed medicinal properties. The species has disappeared from most of Africa; its cousin the black rhino *Diceros bicornis* has fared even worse, and the few thousand which have survived human predation are confined to the best-protected areas. *(Paterson-Jones)*

Fastest of all the mammals of Africa is the cheetah *Acinonyx jubatus* **(above)**, which can attain speeds of up to 100 km/h and more in short, explosive bursts. This beautiful (and now seriously endangered) hunter of the grassland plains, though, has little stamina and can sustain the chase for only a few hundred metres. It first stalks its prey, and then sprints in for the kill. *(Bailey)*

One of the most common animals of the grasslands is the impala *Aepyceros melampus* **(opposite)** – so common in places that game-viewers tend to pass it by. But these antelope are worth more than a second glance: they are renowned for their spectacular leaps over bushy terrain when alarmed, and for the graceful, almost ballet-like choreography of a herd in flight. Only the males have horns, which are lyre-shaped. *(Du Toit)*

Among the harshest of all environments is the Namib Desert, whose high dunes cover a desolate segment of Africa's south-western coastal belt. This grasshopper **(above)** feeds on minute particles of plant matter blown in by the hot winds, and is itself a food source for small vertebrates. *(Dennis)*

Opposite top, centre and bottom: A Peringuey's adder *Bitis peringueyi* burrows into the dune, where it will wait in ambush for passing prey. It is also known as the sidewinder snake for the way it moves laterally, with most of its body kept above the hot sand. *(Dennis)*

Placid though it may appear, the buffalo *Syncerus caffer* **(below)** is in fact one of the most dangerous of African game animals; lone males, which are often to be found lurking near a stream or waterhole, can be especially irritable. The species as a whole, though, is highly gregarious, congregating in herds of anything from a hundred to several thousand. *(Dennis)*

The canine teeth of the warthog *Phacochoerus aethiopicus* **(opposite)** develop into long, curved tusks that serve as useful weapons. Families of this rather ugly but nonetheless appealing, even comical, animal can often be seen tripping along, in strict file, with tails held erect. *(De la Harpe)*

Africa's only resident penguin is the jackass *Spheniscus demersus* **(above)**, the common name derived from its harsh, donkey-like call (heard mainly at night). The bird breeds almost exclusively on the southern offshore islands, though at least two shore colonies have been established – one of them at Boulders, a popular Cape Peninsula beach area. *(Du Toit)*

Spurwing geese *Plectropterus gambensis* **(opposite, top)** feed in the marshlands, floodplains and rivers of Africa. In flight they take position in either staggered lines or 'V' formations. The males are much larger than the females. *(Froneman)*

A familiar presence in freshwater systems throughout the continent is the Egyptian goose *Alopochen aegyptiacus* **(opposite, bottom)**. *(Froneman)*

Mother and child. Lionesses will produce offspring at any time of the year – there is no fixed breeding season – but births within a pride are often synchronized, and a cub suckles from any of the females that are lactating. Prides, which can number up to 30 individuals (though the average is six), are made up of related lionesses and one or perhaps two adult males. *(Balfour)*

Hippos *Hippopotamus amphibius* are found in most of the African continent's freshwater systems, but the populations in many are declining: these bulky herbivores are hunted for their meat and for the ivory of their tusk-like canine teeth. They spend their days in the water or on sandbanks, emerging at night to feed, and often travelling several kilometres in search of grazing. *(Dennis)*

The elephant **(opposite)** is a voracious and destructive feeder: a single adult consumes up to 300 kilograms of grass, shoots, fruit and stripped bark each day, and will topple an entire tree to get to the tender leaves from the crown. The consequences for the natural environment can be devastating, posing a threat both to the herd's well-being and to the survival of other species. In game reserves, where the elephant populations, well protected and healthy, sometimes become too large for the area to which they are confined, surplus animals may have to be culled. *(Balfour)*

One of Africa's lesser-known larger mammals is the puku *Kobus vardoni* **(right)**, a resident of the grassy flatlands near the rivers of south-central Africa (Zambia's Luangwa valley hosts a population of around 40 000). The antelope congregate in herds of up to 30 members, though these are rather unstable, with individuals moving freely between the groups. Only the male pukus have horns. *(Du Toit)*

The bat-eared fox *Otocyon megalotis* **(above)** is an enchanting little animal distinguished by its bushy black tail, rounded back and, its most prominent feature, enormous dish-like ears. These it uses as sound amplifiers to locate termites and beetle larvae that comprise its diet. *(Dennis/SIL)*

Below: Although a member of the canine family, the wild dog is not closely related to the domestic animal: among other differences, it has four toes to each forefoot instead of the usual five. *(Balfour)*

Chacma baboons *Papio cynocephalus ursinus* are excellent parents, highly protective of their young. These primates, which are indigenous to southern Africa, are found mostly in hilly country and along the forest edges, foraging in troops led by a few mature males. They range over a wide area in quest of wild fruits, bulbs, scorpions and, occasionally, small mammals. *(Balfour)*

Feline threesome. Cheetah young **(opposite, top)** are born blind and helpless, and are usually hidden away in thick bush for the first six weeks of their lives. Thereafter they remain with their mother until they can fend for themselves, at which point they move off, the females to establish their own territories, the males to roam the veld alone, in pairs or in small bachelor groups. *(Dennis/SIL)*

Lion cubs **(opposite, bottom)** are often reared by more than one of the pride's females, which enhances their chances of survival. However, the mortality rate tends to be high, the heaviest toll taken by adult male aggression, by starvation and by parasite-born disease. *(Du Toit)*

Below: The female leopard gives birth to between one and three cubs and keeps them with her until they are about a year old. Then, when she is ready to mate again, the maternal ties are decisively and unceremoniously severed. *(Du Toit)*

The quest for food. Black-backed jackals *Canis mesomelas* enjoy a catholic diet, eating anything from young antelope to hares, reptiles, rodents, and berries. This one **(left)** is scavenging a springbok killed by a cheetah. *(Dennis/SIL)*

Below: A steenbok is brought down by a cheetah. The populations of these swift, elegant and surprisingly delicate cats are on the decline: their habitats are shrinking, and they face ever-stiffer competition from the larger and more robust carnivores in their remaining hunting grounds. *(Balfour)*

Popularly held to be the supreme scavengers of the African veld, spotted hyaenas *Crocuta crocuta* are in fact among the most courageous and resourceful of hunters. Above all, they are opportunists, taking whatever food they can find (they have been known to drive lions away from their kill). **Right:** A group drags the remains of a carcass from the water. *(Balfour)*

Overleaf: The majestic beauty of the African elephant, largest of land mammals, and among the longest living: the animal's life-span ranges up to a healthy 70 years. *(Balfour)*

Lions *Panthera leo* feed on a fresh carcass **(above)**. These regal carnivores are the only members of the cat family to live in groups, which confers major advantages during the hunt. Co-operation is especially effective when the prey is a large animal such as a buffalo. *(Balfour)*

The hunt is a complex sequence of prey selection, manœuvre and stalk in which the lioness **(opposite)** plays the major role. The male, though, takes precedence at the carcass. *(Dennis)*

As efficient a hunter in its own way is the handsome caracal *Caracal caracal* **(below)**, found in many parts of Africa and renowned for the speed of its reactions and the strength of its hind legs. *(Dennis)*

Hunters of the night. Owls are famed for their acute eyesight and hearing, and for their ability to fly silently – major assets during the dark hours. Most are also masters of disguise.

The plumage of the African scops owl *Otus senegalensis* **(opposite)** blends beautifully with the tree bark. *(Hes)*

Above: Largest of the family in Africa is the giant eagle owl *Bubo lacteus* , a resident of the drier broadleafed woodlands, thornveld and riverine forests. It is easily recognized by its size, its pink eyelids, and its grunt-like call. *(Knirr)*

The ginger-coloured Pel's fishing owl *Scotopelia peli* **(right)** haunts the hot, humid, tree-lined river banks. It hunts from a perch close to the water, diving in feet first to catch the targeted fish. *(Balfour)*

Previous pages: Bat-eared foxes take their leisure during the heat of the day; twilight will bring out the highly prized harvester termites on which they live. *(Balfour)*

The eyes, ears and nose of the crocodile *Crocodylus niloticus* are so positioned, at the top of its head, that it can stay almost completely submerged for long periods, a threat to and barely noticed by the passing animals and birds. *(Adey)*

A lioness conserves energy during the drowsy, strength-sapping warmth of an African summer's day. The tawny-brown coats of these animals enable them to blend into the long grass of their savanna and bushveld habitat. *(De la Harpe)*

Ancestral to and almost indistinguishable from the household pet – though it is somewhat larger and has longer legs – is the African wild cat *Felis lybica* **(opposite)**, a common enough species about which surprisingly little is known. The animal was first domesticated in Egypt around 6 000 years ago, playing a useful role in keeping granaries free of rats. Interbreeding with its domestic cousins, those which become feral, is threatening its genetic integrity. *(Dennis)*

Almost as poorly studied is the slender, rather handsome serval *Felis serval* **(above)**, a shy, solitary resident of Africa's savanna and moist grassland regions. It feeds mainly on rodents, most especially vlei rats, whose movements it listens for in the long grass. *(Dennis)*

Few hunting techniques are as spectacular as those of the Cape gannet *Morus capensis* **(left)**, a strikingly marked and common enough coastal seabird. To catch its fish prey it plunges down from an impressive height to enter headlong into the ocean. *(Froneman)*

Opposite: Common in both the urban and rural areas of Africa is the soft-cooing Cape turtle dove *Streptopelia capicola*. *(Du Toit)*

Redbilled oxpeckers **(below)** can often be seen clinging to the bodies of giraffe and other large African herbivores – a symbiotic relationship that rewards both parties to the pact. *(Froneman)*

Previous pages: Lioness **(left)** and chacma baboon **(right)** in relaxed mood. *(Balfour)*

The partnership between redbilled oxpecker and impala **(above)** benefits both species: the bird has access to ticks and other tasty organisms; the antelope is ensured of a parasite-free coat. *(Du Toit)*

Opposite: Oxpeckers also hitch a ride on this giraffe. The latter's chief enemy is the lion, to which it is especially vulnerable when drinking: with its legs awkwardly splayed, it is virtually defenceless. Away from water, though, the giraffe is more than capable of holding its own – it is fleet of foot, and can deliver a lethal kick. *(Dennis)*

One of the largest reptiles of its kind is the Namaqua chameleon *Chamaeleo namaquensis* **(opposite)**, which inhabits some of southern Africa's hottest and most desolate areas. It has a rapacious appetite, its long, telescopic tongue reeling in anything up to 200 beetles a day. It also catches such larger prey as locusts, crickets, lizards and even small snakes. *(Dennis)*

Below: The cheetah's relatively light frame, weak jaws and essentially gentle nature put it at a real disadvantage against other, more aggressive predators such as spotted hyaena and lion. Nor can it always keep its kill safe from smaller carnivores such as wild dog, jackal and even vulture. To minimize the chances of conflict over food, it hunts at those times when most of its competitors are inactive, immobilized by the heat of the day. *(Dennis)*

The endearing little ground squirrel *Xerus inauris* **(above)** forages during the day in sparsely covered open areas, using it large bushy tail as a 'parasol' to shield its body from the fierce sun. Most of its diet comprises roots and bulbs, which it cuts through with razor-sharp incisors, but it will also eat fibrous grass, leaves and, occasionally, even termites. This individual is feeding on little devil flowers. *(Du Toit)*

Opposite: A gathering of bright butterflies lends an element of distinction to an otherwise prosaic feature of the African landscape: a dung-heap. *(Du Toit)*

Leopards **(below)** can survive perfectly well on the moisture in their food intake, but will happily drink water when it is available. They are for the most part solitary creatures of the night, coming together only to breed. *(Hes)*

Among the more common herd animals of southern and East Africa is the blue wildebeest *Connochaetes taurinus* **(opposite)**. Aggregations of up to a million and more of these animals set out each year on mass migrations that take them across the plains of the Serengeti and Kenya's Mara River – an unforgettable spectacle. *(Dennis/SIL)*

Overleaf: A brace of shaggy-coated waterbuck *Kobus ellipsiprymnus* in watchful pose. As their name suggests, these antelope are seldom found far from marshland, river or floodplain. *(Balfour)*

World of water. The cavernous yawn of the hippo **(opposite)** is not a sign of sleepiness but either a warning to other males or a prelude to mating. These huge animals are aggressively territorial; rival males will fight each other with terrifying ferocity, sometimes to the death. *(Du Toit)*

One of the odder of Africa's reptiles is the serrated hinged terrapin *Pelusios sinuatus* **(above)** which, when threatened, brings its domed, hinged shell down to cover its head. *(Dennis)*

The African jacana *Actophilornis africanus* **(below)** is uniquely adapted to its watery habitat, its elongated toes and nails enabling it to walk on floating vegetation. *(Tarboton)*

The wader-like doublebanded courser *Rhinoptilus africanus* **(above)**, is a resident of the more arid regions, those embracing semi-desert and dry grassland. Its cryptic coloration provides excellent camouflage in its stony or sandy environment. *(Tarboton)*

Right: The nucleus of a lion pride comprises related females and their offspring. Juvenile males are evicted, by their adult rivals, when they are three years old and move away to form youthful coalitions. On maturity a few years later, they are ready to take on a pride of their own – a period often marked by violent and sometimes lethal confrontations between rivals. *(Balfour)*

Overleaf: Today the prehistoric rhino, a relic of those ancient times when 'mega-herbivores' flourished, faces extinction. However, strenuous and promising efforts have been made to preserve the species, most notably by the conservationists of South Africa's Hluhluwe-Umfolozi Park. *(Hes)*

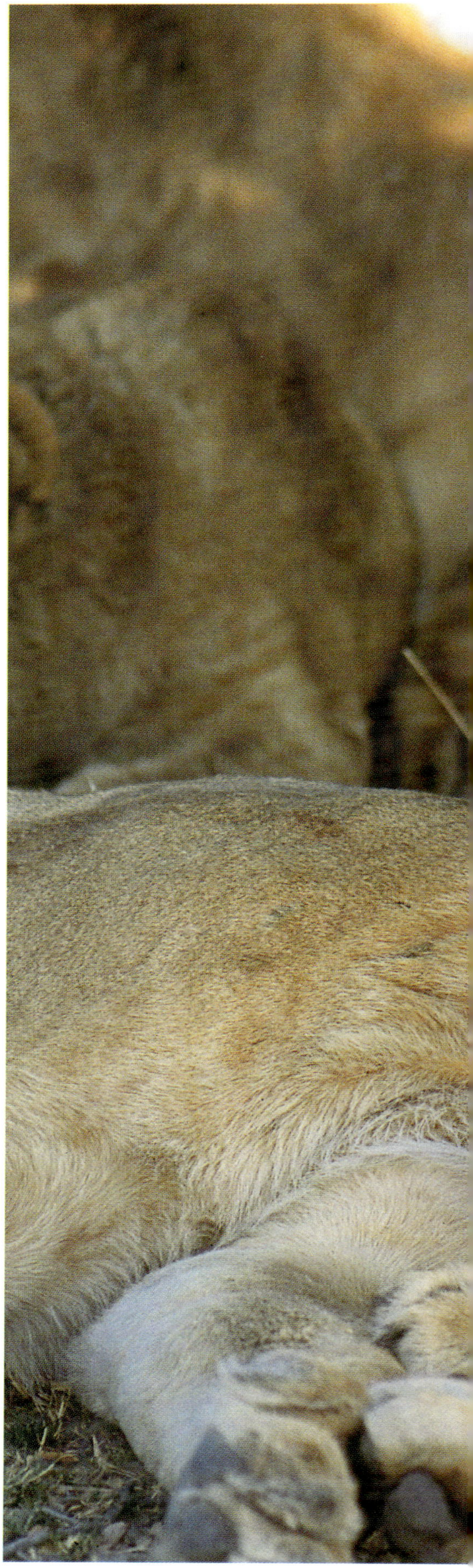